Hilângelo Vieira Barros
João Amaral

Motorbike accidents involving teenagers

Hilângelo Vieira Barros
João Amaral

Motorbike accidents involving teenagers

Epidemiological study in a rural area of north-eastern Brazil

ScienciaScripts

Cover image: www.ingimage.com

This book is a translation from the original published under ISBN 978-613-9-70818-5.

Publisher:
Sciencia Scripts
is a trademark of
Dodo Books Indian Ocean Ltd. and OmniScriptum S.R.L publishing group

120 High Road, East Finchley, London, N2 9ED, United Kingdom
Str. Armeneasca 28/1, office 1, Chisinau MD-2012, Republic of Moldova, Europe
Printed at: see last page
ISBN: 978-620-7-95012-6

ACKNOWLEDGEMENTS

To my family for always being by my side, strengthening me with love in every moment.

To my supervisor Dr João Joaquim Freitas do Amaral for all his support during this course.

To the professors of the Professional Master's Programme in Child and Adolescent Health.

To the course coordinator Prof Dr Maria Veraci Oliveira Queiroz for giving us this unique opportunity.

To Professor Dr Rivelilson Mendes de Freitas for taking the time to answer my questions.

To Professor MSc. Karla Deisy Morais Borges for all her support since the beginning of the master's programme and as a personal co-supervisor.

To Mary Anne Cavalcanti Saraiva Matos for taking the time to advise us on various issues whenever possible.

To my friend Francilena Ribeiro Bessa for letting me know about the course a few days before enrolment closed, and to the rest of my colleagues who gave me so much encouragement on this journey.

To Maria Guedes Lima de Brito ("Dona Maria") who, if it hadn't been for her kindness in enrolling me on the course, would not have been able to complete it.

To all the employees of the Municipal Traffic Department of the municipality of Quixadá for all their support and dedication so that everything could happen as planned in the research.

To the motorcyclists who took part in the study, especially for allowing the research and continuing until its conclusion.

SUMMARY

The high rates of traffic accidents go hand in hand with the growth of the vehicle fleet, and are considered "epidemics" that currently exist all over the world, and have been included on the public health agenda along with morbidity and mortality from external causes. Reducing accidents is one of the biggest challenges for public health, and studies are needed to understand the true magnitude of the problem and the distribution of causes in order to prevent and promote safety. The aim of this study was to investigate the epidemiology of adolescent motorcyclists in a rural area of north-eastern Brazil between January 2008 and July 2010. It was characterised as a retrospective study with an interventionist approach involving 100 adolescent motorcyclists who answered two semi-structured questionnaires during blitz's in strategic locations in Quixadá between April and May 2012. The ethical principles of CNS Resolution No. 196/96 on research with human beings were taken into account. The results of the demographic profile indicated a prevalence of males (92%), aged between 14 and 17 (52%), attending secondary school (86%), mostly single (71%), with a family income of between one and three minimum wages (53%), using motorbikes for work (61%) and predominantly working as motorbike taxis (43%). With regard to motorbike use, only 73% were not licensed, of these 70.4% had a licence at the time of the approach, 55.6% had been driving for between 1 and 12 months (type A), 98% said they did not know the Brazilian Traffic Code, 87% did not wear a helmet, 61% had never worn a helmet when riding, 65% did not practice defensive driving, 91% had their motorbike serviced, 60% had already been involved in accidents, 98% said they did not fully know Law 11.705 ("Dry Law"), 56% say they don't use a motorbike under the influence of alcohol, 72% report having ridden with a drunk motorcyclist, 97% say they have never taken part in educational programmes, 100% of those who have taken part say they took place at school, 90% say there is no traffic education in Quixadá and 93% say they have never been approached about traffic education. This study was able to identify the demographic profile of motorcyclists, as well as possible factors related to accidents involving this population. As a result, I concluded that this topic should be explored in greater depth by other researchers, so that new projects can be set up and we can subsequently arrive at a better diagnosis of the situation experienced by this population with regard to traffic and consequently intervene in the factors most related to accidents, in order to achieve better results in reducing these problems.

Keywords: Traffic accidents; Adolescent health; Public health.

SUMMARY

CHAPTER 1

INTRODUCTION

The vehicle was created to facilitate travelling and interaction between individuals and groups. The vehicle fleet has increased, both due to population growth and the popularisation of this item; however, traffic engineering has not kept up with this growth (LABIAK et aL, 2008).

According to the National Traffic Department (DENATRAN, 2010), the Brazilian fleet has grown by 119% in ten years, reaching a total of 64,817,974 million vehicles in December 2010. Considering the 2010 Census of the Brazilian Institute of Geography and Statistics (IBGE), which indicates that the Brazilian population is 190.732 million, the country has an average of one car for every 2.94 inhabitants. Ceará is the third largest state in the Northeast with 1,711,998 million.

The use of motorbikes is now one of the main reasons for the exorbitant growth in the global transport fleet.

In Brazil alone, there are 13,950,448 million motorbikes, equivalent to 21.5 per cent of all vehicles. In Ceará, the motorbike fleet already exceeds the car fleet (DENATRAN, 2010).

According to data from the Ceará Traffic Department (DETRAN-CE, 2010), from 2004 to 2010, the growth in the motorbike fleet in the state of Ceará was 128.4 per cent, from 282,826 to 645,998.

In Brazil, there were 24,000 deaths in 1997 out of 40,000 accidents involving two-wheeled vehicles, and hospitalisation data by cause group shows that in 1998, 15,232 motorcyclists were hospitalised as victims of transport accidents, while in 2004 this figure rose to 27,388, an increase of 79.8% (SANTOS et aL, 2008).

According to DETRAN-CE (2012), in 2011 there were 26,525 accidents, of which 2,091 resulted in death, of which 761 (36.39%) were motorcyclists. In the municipality of Quixadá, in the same period, there were 103 accidents with 30 (29.1%) deaths.

Therefore, with the disorganised growth in morbidity and mortality rates caused by motorbike accidents throughout the country, it is necessary to create short and medium-term resolving strategies that are capable of reversing this evil.

Brazil has developed important initiatives on the subject, such as the CTB (Brazilian Traffic Code) in 1998, one of the factors responsible for the reduction in mortality rates from transport accidents in the country (SOUZA et al., 2007).

In general, health problems can be attributed to the numerical level of quality of public policies in a given country in terms of morbidity and mortality profiles, as well as the care provided to the population at risk. The WHO calls for stricter policies in order to improve living conditions through agreements made at regular conferences.

The National Policy for the Reduction of Accidents and Violence was therefore enacted in 2001. To implement it, the Ministry of Health's Health Surveillance Secretariat set up the National Accident and Violence Prevention Network in 2004. In turn, the National Agenda for Surveillance, Prevention and Control of Accidents and Violence was approved in 2005 (SANTOS et al., 2008).

Reducing accidents is one of the biggest challenges for public health, and studies need to be carried out to understand the true magnitude of the problem and the distribution of causes in order to prevent and promote safety (CAIXETA et aL, 2010).

Today's negative accident rates can be reduced by a "simple" preventative measure, but this becomes "impossible" when you realise that there are many factors involved in the system, such as: education, training, revision of laws, educational measures, etc. Therefore, how can we solve this problem without the population wanting to learn/change, and how will they learn when professionals are limited by a system that is practically punitive?

However, it is necessary to reorient the training of health professionals so that they can be subjects of the educational process in order to respond to the new demands imposed by traffic victims (VIEIRA et al., 2010).

In this way, educational processes need to be seen not only in terms of the possibility of generating knowledge, but above all in terms of the human dimension and improving people's quality of life and health (JOMARetal.,2011).

This perception is evidenced by the morbidity and mortality data, the need to restructure the roads, the inefficiency of repressing offences and, above all, the effectiveness of educational programmes that raise awareness among the population about risk factor prevention (VIEIRA et aL, 2010).

Today, the subject of education has become more common in the real situation faced by transport users, as well as the bodies responsible for the safety of this population. For many years, this same education was hidden in the existing codes governing traffic behaviour.

Desirable behaviour in traffic is the result of good group education, which will also be reflected in other areas of life, indicating that mutual respect and acceptance of other people's limitations will promote safety and harmony in the interaction between people (DETRAN-PR, 2006).

The stage of human life with the highest risk of death from violent causes is between the ages of 10 and 20, i.e. adolescence. Traffic accidents are one of the main reasons for deaths among adolescents (DE SOUSA; ALMEIDA, 2008).

The high rates of accidents involving motorbikes throughout Brazil have sparked interest at municipal level in studying the epidemiology of motorbike accidents involving adolescents in order to develop educational strategies.

The municipality of Quixadá is no different, with a population of 80,605 inhabitants according to IBGE - CENSO 2010 and a fleet of 17,183 vehicles, totalling one vehicle for every 4.69 inhabitants. Motorbikes lead the way in terms of numbers, accounting for 9,073 (52.8%) of the total (DETRAN-CE 2010).

It is believed that this knowledge can serve as a basis for better enlightening the community about the risks to which they are exposed and the importance of safe behaviour in traffic.

There are a number of transport accidents and in greater proportions there are those caused by motorbikes, most of which are teenagers, unlicensed, drinkers and without the use of PPE (Personal Protective Equipment).

These are the reasons that explain the real interest in developing this study, together with the records found in the DMT (Municipal Traffic Department) of Quixadá, in which there is a high prevalence of traffic accidents involving, above all, the use of motorbikes by adolescents.

This whole problem must, above all, be backed by a political project that aims to considerably reduce the entire context of policies aimed at resolving the problems caused by accidents and that are not just "welfare" but preventive, as well as resolving and continuous.

It is with this idea in mind that we aim to study the epidemiology of motorbike accidents involving teenagers in the municipality of Quixadá from 2008 to 2010, using data obtained from the competent body (DMT of Quixadá), in order to approach the population at strategic points in the city, through educational *blitzes*, and explain the real importance of reversing this situation.

CHAPTER 2

LITERATURE REVIEW

The high rates of road traffic accidents go hand in hand with the growth in the vehicle fleet, which is considered an "epidemic" around the world and has been included on the public health agenda along with morbidity and mortality from external causes.

According to data from RENAINF (Registro Nacional de Infrações de Trânsito), speeding is the most common offence committed by vehicles outside their state of registration; from January 2004, when the system began to be implemented in the country, until July 2010, 22,780,514 traffic offences were recorded, of which 9,834,097 were for speeding (DENATRAN, 2010).

According to Labiak et al. (2008) human behaviour, and not just traffic engineering problems or vehicle deficiencies, has been identified as the main cause of traffic accidents. Among young people, a lack of adaptation to traffic and age-related behaviour, such as impulsiveness, daring and the search for novelty, are factors associated with a higher risk of traffic accidents. And it is these behaviours that have generated huge health concerns in general, due to their great impact on society. Some thematic publications on accidents and violence show that traffic accidents have an impact on the physical, emotional and social health of individuals and families, as well as causing serious damage to the country's economy (VIEIRA et al., 2010).

These incalculable impacts only reinforce how serious the situation is for the victims and others involved in traffic accidents, and that today there is already greater concern about these negative rates.

Traffic accidents are currently a serious public health problem because they are accompanied by a high morbidity and mortality rate (ANDRADE et al., 2009).

This same problem has been affecting many people around the world and consequently causing serious suffering, jeopardising the health of the population to a considerable extent.

The high frequency and severity of road traffic accidents are among the main negative effects on the population's health conditions. It is estimated that between 20 and 50 million people are injured every year. In fact, road traffic injuries account for 2.1 per cent of all deaths and 23 per cent of deaths from external causes recorded annually worldwide, and are among the leading causes of death among young men aged between 15 and 44 (OLIVEIRA; MOTA; COSTA, 2008).

The World Health Organisation (WHO; 2008) points out that the relevance of the issue is due to its magnitude and also because traffic is part of people's daily routine, which is becoming more complex and dangerous every day.

Traffic accidents make up a considerable percentage of external causes, since morbidity and mortality rates exceed even other problems that were previously seen as priorities.

According to the WHO, the number of people who die each year as a result of road traffic accidents accounts for 25 per cent of all external causes of death on the planet. Estimates point to a growing trend in these figures, which are expected to increase by 40 per cent by 2030 if effective preventive measures are not adopted, especially in developing countries (JOMAR et al, 2011).

Andrade et al. (2009) state that this type of accident is among the most common external causes, with a high percentage of hospitalisations, as well as high hospital costs, material losses, social security costs and great suffering for the victims and their families. Mortality from these causes has reached such proportions that it has had an impact on the population's life expectancy, especially among men.

Traffic accidents represent a significant social burden, not only because of the deaths and sequelae they cause, but also because they burden society with direct and indirect costs, estimated at 1 to 2 % of the Gross Domestic Product (GDP) per year (JOMAR et al., 2011).

Since then, they have accounted for 15 per cent of deaths in the country, second only to diseases of the circulatory system. According to data from the National Health Council (CNS), there are 120,000 deaths a year in Brazil; these figures correspond to the total number of people killed in historical conflicts, such as the Vietnam and Gulf wars. There are 630,000 hospitalisations and a financial cost to the SUS of at least R$ 351 million a year. The length of these hospitalisations varies between five and eight days, which suggests that serious injuries have occurred (IMPERATORI; LOPES, 2009).

Among the main forms of transport involved in this data, motorbikes stand out, since their use has been growing at alarming proportions and consequently their problems have become worldwide.

The problem, however, is that in the face of this growth, the recklessness of drivers, lack of enforcement and lack of infrastructure for this type of vehicle, there are consequences such as an increase in traffic accident statistics (DETRAN-CE, 2010).

Silva et al. (2011) state that the mortality rate for this category has risen dramatically since the mid-1990s. The mortality rate for motorcyclists rose by 875% between 1996 (0.4/100,000 inhabitants) and 2006 (3.9/100,000 inhabitants).

Motorbikes have been gaining huge ground because they are cheaper than other forms of transport, consume less fuel and are easy to manoeuvre for those who "don't have time to waste", usually motorbike taxis and *"motoboys".*

The vulnerability of motorbike users has been demonstrated by the number of accidents involving this type of vehicle. In the impact of motorbike accidents, there is often an unequal collision with larger vehicles. On the other hand, the motorcyclist does not have the structure of the vehicle to protect him, absorbing all the energy of the impact and commonly being thrown a distance (OLIVEIRA; SOUSA, 2006).

In this context, it is understood that motorcyclists, because they lack adequate protection, are the main cause of these events. The number of hospitalisations for these causes in Brazil and around the world is alarming.

According to the newspaper Diário do Nordeste (2010), traffic accidents on Ceará's state motorways increased by 10.99 per cent between February 2009 (191 accidents) and the same period in 2010 (212 accidents).

Because of all these problems, it became necessary to organise the country so that there were rules and laws of conduct that covered the entire national territory in terms of traffic, and so the Brazilian Traffic Code was created.

The CTB is an efficient instrument, but if the law is not rigorously enforced by the responsible authorities, drivers will continue to break traffic laws without social awareness, and so punishment becomes a means of curbing this type of behaviour. There is therefore a need to step up enforcement and apply the appropriate punishments when necessary (DETRAN-PR, 2006).

Unfortunately, this code has not been working, since for many people and in many places it is seen as a factor of political interests, especially in small towns where the majority of offenders have political influence. From then on, new ideas have been created to integrate with the existing ones in the system to cover gaps in the programmes.

In 1998, the Ministry of Health set up a technical committee with the aim of diagnosing and proposing specific actions for the sector, bearing in mind that the mortality and morbidity profile of the Brazilian population is marked much more by conditions, situations and lifestyles than by traditional illnesses, and that violence and accidents deserve as much attention as AIDS (Acquired Immune Deficiency Syndrome), cancer and cardiovascular diseases (SANTOS et aL, 2008).

In general, health problems can be attributed to the numerical level of quality of public policies in a given country in terms of morbidity and mortality profiles, as well as the care provided to the population at risk. The WHO calls for stricter policies in order to improve living conditions through agreements made at regular conferences.

Of all the policies already in place to minimise the high rates of these diseases, we can cite the use of alcohol as a recent priority in the country due to the binomial factor that exists between driving and the use of alcohol.

In this context, it is known that a significant proportion of road traffic accidents can be attributed to excessive alcohol consumption. It is estimated that 70 per cent of fatal traffic accidents are related to the use and abuse of alcoholic beverages (JOMAR et al, 2011).

This evidence influenced the Brazilian Congress to implement Law No. 11.705 in 2008, which reduces the permitted blood alcohol level to zero, increases the administrative penalty and criminalises drivers who drive with 0.6 dcg or more of alcohol per litre of blood (MOURA et aL, 2009).

As well as reducing the use of alcohol, other public policy priorities have shown progress in decision-making in order to achieve a reduction in the number of health problems affecting the population. Discussions on traffic education are now more frequent.

Reducing accidents is one of the biggest challenges for public health, and studies need to be carried out to understand the true magnitude of the problem and the distribution of causes in order to prevent and promote safety (CAIXETA et aL, 2010).

According to Paulino (2010), Brazil's first Traffic Code dates back to 1941 and, throughout its 12 chapters, at no point did it allude to the subject of education. The first references were made in the National Traffic Code, established by Law no. 5.108 of 21 September 1966, which made only passing mention of the requirement for a representative from the Ministry of Education and Culture to sit on the National Traffic Council, the provision of traffic education campaigns and, finally, the dissemination of traffic notions in the country's primary and secondary schools.

Dílson de Souza Almeida, a technical advisor to CONTRAN (the National Traffic Council), gave a lecture in 1999, i.e. just after the current Brazilian Traffic Code (in force since 22 January 1998) came into being, and observed, quite sagely, that in the CNT (the 1966 one): The word education appears twice; the term educational campaigns, four times; the word learning, twice. Thus, the subject of education appears a total of eight times, which represents just over 6 per cent, taking into account the 131 articles of the Law. As for the CTB, the astute scholar noticed that it contains the word education twenty times: The word education can be read twenty-eight times, in addition to a further 13 words and related terms (learning, educational campaign, specialisation, teaching level, teaching curriculum, interdisciplinary curriculum, public school, etc.) which appear twenty-one times. The subject is therefore covered forty-nine times, which represents 15 per cent of the 341 articles in the Law (PAULINO, 2010).

We can see, then, that various measures have been taken to organise public policy in the area of accident and violence prevention; however, the most important principle has been forgotten, which is educating the population about traffic and, as a priority, schoolchildren, who will soon be the actors in this issue.

Unfortunately, both professionals and politicians do not value what is most effective, which would be traffic education. The main objective of primary care policy is to work with the population practically to prevent illnesses, thus promoting campaigns, lectures, information through the media, among others, and if there were a plan for continuous and periodic adherence to this type of care; since the re-education of a population takes a long time and a lot of patience.

Health education can be understood as a set of knowledge and practices aimed at disease prevention and health promotion (JOMAR et al. 2011).

CHAPTER 3

OBJECTIVES

3.1 General Objective

To assess the epidemiology of adolescent motorcyclists in a rural area of north-eastern Brazil between January 2008 and July 2010.

3.2 Specific objectives

- To determine the number of motorbike accidents between January 2008 and July 2010;
- To describe the socio-demographic characteristics of adolescent motorcyclists;
- To characterise motorcyclists in terms of the legality, prevention and risks of their behaviour in traffic;
- Identify actions taken by teenagers that reveal their traffic education;
- Describe how long they have been driving and their knowledge of the Brazilian Traffic Code.

CHAPTER 4

METHODOLOGY

4.1 Delineation

This is a retrospective study with an interventionist approach, which seeks to examine past data and samples, which will be presented to the participants at a later stage, through six *blitzes* in pre-determined locations with the application of questionnaires relating to socio-economic standards and the use of motorbikes by adolescents.

4.2 Place and period of the study

The place selected for the study was the city of Quixadá, a Brazilian municipality in the state of Ceará, located in the sertão mesoregion of Ceará. Its population is 80,604, predominantly urban (67.3%) and female (50.3% of the total).

The municipality is divided into 13 units: The Head Office and 12 other districts. In 2005, the municipality had 145 primary and secondary schools, 16 of which were private (11%). The school enrolment rate is 100% for primary education and 42.86% for secondary education. It has five higher education institutions and one technical school.

FIGURE 1 - Geographical location of the city of Quixadá - Map of the state of Ceará.
SOURCE: Image from the Google Maps website.

Epidemiological data on motorbike accidents involving adolescents from January 2008 to July 2010 was collected from the Quixadá DMT archives in November 2011.

Data collection on socio-economic status and motorbike use by adolescents was carried out in August 2012 in the city of Quixadá.

4.3 Study population

The study population consisted of 100 adolescent motorcyclists living and resident in Quixadá. The epidemiological assessment used data on motorbike accidents from the Quixadá DMT from January 2008 to July 2010. The cross-sectional evaluation used data obtained from questionnaires applied at the time of the educational *blitz*.

The sample size was calculated based on an estimate of the population mean, with a sampling error of 5%, 95% confidence level and minimum percentage of 7% using data on the total motorbike fleet (9073 motorbikes) in Quixadá (DENATRAN, 2010).

$$n = \left(\frac{Z_{\alpha/2} \cdot \sigma}{E} \right)^2$$

n= Number of individuals in the sample

$Z\alpha/2$= Critical value corresponding to the desired degree of confidence

σ= Population standard deviation of the variable studied E = Margin of error or MAXIMUM ESTIMATE ERROR

4.4 Eligibility criteria

The primary study included all adolescents aged between 10 and 20, according to the classification of the Ministry of Health (Ordinance 980 of 21/12/1989) and the Brazilian Society of Paediatrics (1998), who use motorbikes as a means of transport. Only adolescents who were driving a motorbike at the time of the *Blitz* were included in the cross-sectional study.

4.5 Data collection

The data was collected at two different times: The first involved an epidemiological survey of accidents (total number of accidents, age group involved in the accidents, type of transport involved in the accidents and the locations where the events occurred) provided by the DMT in the city of Quixadá from January 2008 to July 2010 (APPENDIX A). These were used to educate the research subjects at the time of the *blitz* and to guide the main locations, as well as the number of *blitz events;* these were divided according to the importance of the data: Two blitz were organised in the centre and only one blitz was carried out in each of the other neighbourhoods (Campo Novo, São João, Campo Velho and Alto São Francisco), totalling 06 blitz in the period from April to May 2012, all of which took place on weekends.

In a second phase, six educational *blitzes* were carried out. During the approach, each volunteer participant was given a Free and Informed Consent Form (APPENDIX C), in which the motorcyclist was informed about the research and signed (if the teenager was a minor, an adult guardian had to sign another form - APPENDIX D) consenting to all the information contained therein, guaranteeing the preservation of their identity, as well as withdrawing from the research when requested.

Two pre-structured questionnaires were used as data collection instruments: the first consisted of questions about the demographic profile (APPENDIX A) of the participants, such as age, gender, marital status and family income. The second consisted of questions relating to motorbike use (APPENDIX B) such as: driving licence, helmet use, alcohol use, motorbike overhaul, knowledge of the traffic code, history of involvement in accidents and participation in educational *blitzes*.

4.6 Data processing and analysis

The variables acquired through the information provided by the Quixadá DMT and the two questionnaires obtained by the researcher were used to analyse the data. They were then tabulated and analysed using the *EPIINFO* software.

4.7 Ethical aspects

The research was sent to the Research Ethics Committee (CEP) and only started after approval (APPENDIX B), with the authorisation of the head of the Quixadá Municipal Traffic Department (APPENDIX E) and the signing of the Free and Informed Consent Form by the parents or guardians (APPENDIX D) and the subjects (APPENDIX C).

The confidentiality of the identity and information collected from each participant was guaranteed, as well as the freedom to withdraw at any stage of the research without any harm to the volunteer.

The research was classified as Approved by the Research Ethics Committee (CEP) of UECE, under protocol no. 11042183 3; according to the regulatory norms contained in Resolution 196/96, which guides the ethics of research with human beings.

CHAPTER 5

ANALYSING THE RESULTS

According to the data collected from the Quixadá DMT archives (ANNEX A), 1,484 transport accidents were reported between January 2008 and July 2010. The most affected age group was between 17 and 24 years old, with 533 (35.9%) accidents. Motorcycles prevail as the transport most involved in accidents, with a total of 1,342 (90.4%) and 58.8% (873 accidents) occurring in 5 of the 74 localities (neighbourhoods and districts) studied.

The socio-demographic profile was carried out using a questionnaire (APPENDIX A) in order to get to know our study population and thus be able to draw up preventive strategies according to the situation they are experiencing.

TABLE 01 - Distribution of the sample in terms of frequency and percentage of sociodemographic data of the motorcyclists interviewed. Quixadá/CE, 2012. n = 100.

Variables	Frequency (f)	Percentage (%)
Age group		
From 10 to 13 years old	05	5,0
From 14 to 17 years old	**52**	**52,0**
18 to 20 years old	43	43,0
Total	**100**	**100,0**
Sex		
Male	**92**	**92,0**
Female	08	8,0
Total	**100**	**100,0**
Education (completed)		
Primary school	03	3,0
High school	**86**	**86,0**
Higher education	11	11,0
Total	**100**	**100,0**
Marital status		
Single	**71**	**71,0**
Married	23	23,0
Divorced	06	6,0
Total	**100**	**100,0**
Family income		
From one to three minimum wages	**53**	**53,0**
From four to six minimum wages	39	39,0
More than six minimum wages	08	8,0
TOTAL	**100**	**100,0**
Type of transport used for work		
Cars	09	9,0
Motorbike	**61**	**61,0**

Others	28	28,0
Total	**100**	**100,0**
Profession that requires a motorbike		
Mototaxi	**43**	**43,0**
Motoboy	10	10,0
Don't work	21	21,0
Total	**100**	**100,0**

As shown in Table 1, 52% of the motorcyclists interviewed were aged between 14 and 17, predominantly male (92%), 86% were in secondary school, 71% were single, 53% had a family income of between one and three minimum wages, 61% said they used their motorbike for work and 43% said they worked as a mototaxi and/or *motoboy.*

The instrument on the relationship between adolescents and motorbike use (APPENDIX B) enabled a general analysis of the frequency with which events occurred, allowing important data to be obtained and later used in preventive strategies. The data from the above-mentioned instrument was divided into tables according to its classification (driving licence, traffic safety, alcohol use and traffic education).

TABLE 2 - Distribution of adolescents in terms of frequency and percentage of data relating to motorbike driving licence (type A). Quixadá/CE, 2012.

Variables	Frequency (f)	Percentage (%)
They have a licence		
Yes	27	27,0
No	**73**	**73,0**
Total	**100**	**100,0**
Driving licence at the time of approach		
Yes	**19**	**70,4**
No	08	29,6
Total	**27**	**100,0**
Time taken to obtain a licence		
From 01 to 12 months	**15**	**55,6**
From 13 to 24 months	09	33,3
25 to more months	03	11,1
Total	**27**	**100,0**

Table 2 shows that only 27% of the drivers were qualified to ride a motorbike; of these, 70.4% had a driving licence at the time they were approached and 55.6% had held their licence (type A) for between one and 12 months.

TABLE 3 - Distribution of adolescents in terms of frequency and percentage of data relating to traffic safety. Quixadá/CE, 2012. n = 100.

Variables	Frequency (f)	Percentage (%)
Fully acquainted with the Brazilian Traffic Code		

Yes	02	2,0
No	**98**	**98,0**
Total	**100**	**100,0**
Helmet present when approached		
Yes	13	13,0
No	**87**	**87,0**
Total	**100**	**100,0**
Frequency of helmet use		
Always	06	06,0
Sometimes	33	33,0
Never	**61**	**61,0**
Total	**100**	**100,0**
Practise defensive driving		
Yes	35	35,0
No	**65**	**65,0**
Total	**100**	**100,0**
Have your motorbike serviced at the indicated times		
Yes	**91**	**91,0**
No	09	9,0
Total	**100**	**100,0**
Ever been involved in an accident		
Yes	**60**	**60,0**
No	40	40,0
Total	**100**	**100,0**

Table 3 refers to traffic safety, where 98% said they didn't know the CBT, 87% didn't use helmets as a preventive and safe way of avoiding possible injuries in accidents, 61% said they never wore a helmet when riding a motorbike, 65% never practised defensive driving, 91% had their motorbike serviced on schedule and 61% said they had already been involved in accidents.

TABLE 4 - Distribution of adolescents with regard to frequency and percentage of data on the use of alcoholic beverages. Quixadá/CE, 2012. n = 100.

Variables	**Frequency (f)**	**Percentage (%)**
Complete knowledge of Law 11.705 "Dry Law"		
Yes	02	2,0
No	**98**	**98,0**
Total	**100**	**100,0**
Drinking alcohol while riding a motorbike		
Yes	44	44,0
No	**56**	**56,0**
Total	**100**	**100,0**
Rides or has ridden with an intoxicated motorcyclist		
Yes	**72**	**72,0**
No	28	28,0
Total	**100**	**100,0**

Table 4, on the use of alcoholic beverages, shows that 98 per cent of the motorcyclists interviewed said they were completely unaware of Law 11.705 ("Dry Law"), 56 per cent did not use any alcoholic beverages when riding a motorbike and 72 per cent said they were riding or had ridden as passengers where the rider had used alcoholic beverages.

TABLE 5 - Distribution of adolescents in terms of frequency and percentage of data relating to traffic education. Quixadá/CE, 2012.

Variables	Frequency (f)	Percentage (%)
Have you ever taken part in a traffic education programme?		
Yes	03	3,0
No	**97**	**97,0**
Total	**100**	**100,0**
If so, where did it take place?		
At school	03	100,0
Total	**03**	**100,0**
There's education in Quixadá's traffic		
Yes	10	10,0
No	90	90,0
Total	**100**	**100,0**
Have you ever been approached by a traffic officer about education?		
Yes	07	7,0
No	**93**	**93,0**
Total	**100**	**100,0**

Table 5, referring to traffic education, shows us that 97% said they had never taken part in traffic education programmes in Quixadá, of those who had taken part in these programmes 100% said they had taken place at school, 10% said there was traffic education in Quixadá and 07% said they had already been approached about traffic education in Quixadá.

CHAPTER 6

DISCUSSION

The results found in Table 1 are consistent with other studies on motorbike accidents involving adolescents.

Analysing the data showed the prevalence of males as the main users of motorbikes, which is all the more worrying when we realise that the majority are teenagers under the age of 18. Therefore, they should not be in charge of riding such an incident and prevalent form of transport, with negative data on deaths and injuries resulting from its misuse.

Unfortunately, the population is involved in these attitudes (directly or indirectly) by teaching and/or giving their transport to teenagers even though they know it's irregular.

According to Anjos et al (2007), "the qualification of young drivers of motor vehicles represents a real rite of passage in the modern world". It is therefore up to those responsible to encourage them to practise legality, limiting them to using vehicles from the age of 18, as established by law. Caixeta (2010) states that the predominance of males is probably due to the greater exposure of men in traffic, socially and culturally influenced so that they take greater risks when driving vehicles. And Pinto and Witt (2008) attribute it to their greater involvement in the labour market, leading to greater exposure as a result of professional activity and commuting to work. We can therefore see that these factors are linked to the occurrence of these events.

The level of schooling becomes extremely important in terms of the applicability of preventive actions, which can produce results in the short and medium term, as the study indicates the need for immediate interventions on traffic in schools, especially at secondary level, without forgetting the other levels of schooling. Pinto and Witt (2008) state that the level of education found is similar to that of other studies, in which the majority had a level of education no higher than secondary school and, according to Santos et aL (2008), educational interventions will therefore be necessary at primary, secondary and higher education levels.

Adolescents, most of whom are single, need to work in order to be financially independent or even to take responsibility for their family's daily expenses. This shows the reality experienced by many wage earners, especially those in the countryside, who practically only have income from public office and pensions.

The motorbike has become indispensable both because it's easy to acquire (low cost) and because it guarantees autonomous services, as is the case with the growing number of mototaxi drivers. This profession demands a lot from the worker and, according to Veronese and Oliveira

(2006), requires agility and precision that are often compared to urgency and competition. The urgency required for deliveries demands high speeds. This is a recognised cause of accidents and is condemned in the Brazilian Traffic Code. Therefore, professionals who use motorbikes for their work need to be included in a preferential lane for preventive events organised by the competent bodies, in this case the Quixadá DMT.

Table 2 shows that the data is negative, since this study showed a predominance of people aged outside the limits for driving licences, a fact seen in the study by Françoso and Coates (2008) when they stated that the high frequency of drivers under 18 years of age and young people without a licence is extremely worrying; and that Benincasa and Rezende (2006 apud ANDRADE et al., 2003) add that the risk factor "lack of driving skills" portrays the reality of the participants' routine, as they are individuals aged up to 18, do not yet have a driving licence, have been driving for a short time and have no knowledge of traffic laws.

It was also possible to observe that of the few teenagers who have a licence, they are irresponsible when driving a motorbike without a licence; this reinforces the importance of greater monitoring by the competent bodies to ensure that they are more interested in complying with traffic laws.

The length of time the interviewees have been driving reveals a minimum degree of experience in traffic; this is described in the study by Liberatti et al (2003), which states that the short time of experience in driving vehicles, associated with such factors (use of alcoholic beverages, speeding, non-use of PPE, etc...) make this group the most at risk of being involved in traffic accidents. However, they have a slight positive growth rate compared to the others, which gives us the idea that there is a greater demand for and approval of driving school courses.

On the other hand, the data in Table 3 shows that drivers are unsafe and at greater risk in traffic.

In their study, Campos et aL (2007) state that the rules of the Traffic Code for this group are not well known and that the population is unaware of their rights and duties as pedestrians and users of the traffic system. This shows us the need for an educational strategy aimed at traffic safety, especially in schools, since children and teenagers "will be the future drivers".

With regard to the use of helmets as a preventative measure, most of the interviewees were not interested. In a study carried out in Los Angeles, United States, Santos et aL (2008) showed that the use of helmets can reduce the number and severity of head and neck injuries, as well as preventing traumatic brain injuries and cervical fractures [...] and observed that motorcyclists who were not wearing helmets at the time of the accident were twice as likely to suffer traumatic brain injuries as those wearing helmets.

In the same vein, none of the interviewees reported on the use of other PPE, making it clear that the obligation was limited to the use of helmets. This shows a lack of knowledge of the law, as it is known that Resolution 203 of the Brazilian Traffic Council, in its art. 1° , states that it is compulsory for motorbike drivers and passengers to wear a helmet when travelling on public roads, and that the Brazilian Motorcyclists' Association publishes a list of twelve commandments for motorcyclists' safety on Brazilian streets and roads, which includes: a helmet approved by INMETRO (National Institute of Metrology, Quality and Technology), trousers and a jacket made of resistant fabric, reinforced boots or shoes and gloves. They are also required to wear leg protectors and anti-glare antennae (SANTOS et aL, 2008).

So we see the need for not only public bodies but also users to fulfil their rights and duties so that there is greater integration between them and consequently a lower rate of accidents and injuries that may occur.

The lack of proper helmet use at the time of the approach reveals a directly proportional relationship with the frequency of helmet use. It is becoming increasingly common not to wear a helmet, especially among young people; they report great discomfort in wearing them, as well as possible fungal diseases due to the humidity provided by sweat when exposed to sunlight. In addition, many compare the use of helmets to the various crimes that exist, most of which are used to protect the face so as to remain unrecognisable and consequently not be arrested.

Some towns in the interior of Ceará use these beliefs and have adopted the non-use of helmets by motorcyclists, mainly because there are so many gun-running crimes, where practically everyone involved uses this article.

The practice of defensive driving was shown in this study not to be seen as important.

Nowadays, preventive drivers must use their reflexes not only for themselves, but above all for others responsible for traffic, because accidents can be avoided both by obeying the law and by always being aware of the events around them, especially other inattentive drivers.

However, this data is predictable due to the "natural behaviour" of this age group, as stated by Anjos et al. (2007) when they say that the challenges added to inexperience in driving vehicles, the lack of knowledge of traffic regulations, the ease of using alcohol, the feeling of invulnerability causing the impression of immortality, the result of a lack of risk awareness". In the case of workers, Caixeta et al. (2010) state that the quest to increase productivity by adopting unsafe measures, such as risky manoeuvres and high speeds, may explain the high occurrence of occupational accidents involving motorbikes.

In these two classes there is recklessness and failure to practice defensive driving due to intrinsic behaviour, which is an individual attitude that comes from the person themselves, and extrinsic factors that vary depending on the use of the motorbike; in the case of work, the need for speed and agility to reach that goal in the shortest possible time.

With regard to having the motorbike serviced at periods pre-established by the manufacturer, we observed the existence of preventive acts, which can show the positive proportion of workers and respect for the importance of maintaining the work tool, thus avoiding possible future accidents related to its maintenance.

According to Debieux et al. (2010), the cause of accidents is multifactorial and is related to a combination of factors, highlighting, among the many, the state of repair of motorbikes as one of the main ones. In another study, motorbike *couriers* also emphasised the importance of motorbike maintenance (VERONESE; OLIVEIRA, 2006).

The involvement of this class in accidents is proportional to the amount of this type of transport. This can be explained by the fact that the majority are workers and constant users of motorbikes, as is the case with mototaxis and *motoboys,* because they are forced to work against the clock. According to Silva et al. (2011) *Motoboys* and motorbike taxis are new forms of work and paid passenger transport that are increasingly used by companies and individuals, and according to Veronese and Oliveira (2006) they are encouraged to compete by their bosses and clients. Therefore, the quest to increase productivity by adopting unsafe measures, such as risky manoeuvres and high speeds, may explain the high occurrence of occupational accidents involving motorbikes (CAIXETA et al., 2010).

Another fact can be observed in relation to experience, as Veronese and Oliveira (2006) add that an inexperienced motorcyclist contributes to the risk of traffic accidents [...]. Excessive experience, on the other hand, can lead them to learn risky ways of driving.

According to Table 4 on full knowledge of Law 11.705 of 19 June 2008 ("Dry Law"), the vast majority are not aware of their rights and duties as citizens, nor are they aware of their responsibilities in traffic. This shows us once again the importance of traffic education in strategic places (schools, squares, etc.) and not just the enforcement of laws and fines. This is because the majority of young people who are about to obtain their driving licence are as willing to drive drunk as they are to use safe alternatives after consuming alcohol (SANTOS et al, 2008).

The use of alcoholic beverages when driving transport seems to be increasingly common and is only increasing every year, which shows us a high number and consequently a greater likelihood of accidents occurring with them in the future. In their research, Vieira et al (2011) state that drinking alcohol slows down reflexes and affects vision, as well as causing a feeling of euphoria that induces the driver to speed and lose control of the motorbike. And it is because of these changes and many others that this rate of drinking and driving is always being discussed around the world, and is treated as one of the main problems to be solved.

Several studies point to a strong relationship between alcohol consumption and traffic accidents [...] This is a regrettable habit, responsible for the loss of many lives, since drunk drivers are four times more likely not to wear a helmet (SANTOS et aL, 2008); they are therefore more vulnerable to fatal accidents. According to Liberatti et aL (2003), drinking alcohol before driving is an important risk factor for traffic accidents.

We can also see that they don't worry about the risk they run by entrusting their lives to a drunk driver when they get on a motorbike.

Table 5 refers to traffic education and, according to Caixeta et aL (2010), schools can offer a good opportunity for intervention, given the importance of education in preventing transport accidents. Involving schools in the development of public policies seems to be a natural way to prevent and reduce traffic accidents in the medium and long term, by monitoring parents and training more aware pedestrians and drivers. This data is showing a return in terms of improvement and quality of life for all those involved in traffic. Bearing in mind that the Brazilian Traffic Code (1997) establishes in its chapter VI, traffic education, in its article 74 it says that traffic education is the right of all and constitutes a priority duty for the components of the National Traffic System; and in article 76 it states that traffic education will be promoted in pre-schools and schools of 1° , 2° and 3° grades, through planning and coordinated actions between the bodies and entities of the National Traffic and Education System, of the union, the states, the federal district and the municipalities, in their respective areas of activity. We have seen that educational measures are important (BACCHIERI; BARROS, 2011). We can see a lack of attention to compliance with these laws, and perhaps a greater guarantee of the quality and quantity of these events is needed for positive responses to appear.

The rate of occurrence of educational programmes in traffic is worrying due to the low occurrence of this event. Anjos et al. (2007) state that educational and preventative work is essential in these cases, as acting on the causes is much easier than on the effects, which are serious traffic accidents and their various sequelae.

Still on the subject of education, Caixeta et al. (2010) state that schools can offer a good opportunity for intervention, given the importance of education in preventing transport accidents and that the Brazilian Traffic Code itself recommends that the union, states, municipalities and the Federal District should interact to promote traffic education, from pre-school to high school.

The survey reveals that the little education that exists was seen with students, which shows us once again the importance of inserting practices aimed at schoolchildren.

According to Vieira et al. (2010), health education is not just about preventing illnesses and diseases, it is a process of qualifying the individual [...]. Ways of preventing health problems are effective through interdisciplinary and intersectoral action (transport, traffic engineering, safety, education and health), aimed at reducing the number of transport accidents, focusing on risk factors such as alcoholism, high speed, drowsiness and tiredness, among others, and reinforcing the notion of danger in the face of offences committed by traffic users.

In the same vein, we can mention a project by the Department of Labour and Entrepreneurship and the State Traffic Department of Piauí to train motorcyclists, covering training in safety, legislation and human relations, including the issue of traffic education (SANTOS et al., 2008); showing us that, in addition to strategies with schoolchildren, we must work on the ongoing training of the main players in the current reality, which are traffic professionals and transport users.

Unfortunately, the competent bodies always seem to take action when there is a negative index of any event and they are looking to improve the results, while the others are forgotten until they also become negative and require intervention. We have to evaluate and seek solutions to the various problems that exist and thus guarantee their continuity, respecting people's rights, because what users say is that there are only approaches aimed at punishment, fines, and the few times that educational intervention takes place is when there is a commemorative day related to traffic.

In their study, Parreira et al. (2012) say that specific prevention campaigns should be publicised and laws strictly enforced. Bacchieri and Barros (2011) state that educational measures are important and widely used, but are not effective, especially when used in isolation. This shows us that there are very active proposals, but that they are not used for various reasons, and that the penalty is a strategy to circumvent the errors in the system, especially when it is seen as a lucrative form of legal collection.

CHAPTER 7

CONCLUSION

This study was able to ascertain the epidemiology of accidents involving adolescent motorcyclists, as well as possible factors related to accidents involving this population.

Motorbike accidents involving teenagers deserve more attention due to the number of these events and education seems to be a good initiative, since the data revealed a need for risk prevention and the application of traffic laws and rules of conduct.

Some actions have been observed with regard to educating teenagers about traffic, but the number and permanence of these initiatives must be intensified and it is extremely important that they start from an early age, even before they become drivers; because knowledge and experience must be taken into account. With this in mind, I concluded that this topic should be explored and deepened by other researchers, so that new projects can be set up and we can subsequently reach a better diagnosis of the situation experienced by this population in terms of traffic and consequently intervene in the factors most related to accidents, in order to have a better result in reducing these injuries.

Educational and preventive work aimed at motorbike drivers should be carried out at pre-established dates and locations, ensuring their permanence and continuity.

The training of the professionals involved and the training of motorcyclists is something that must be applied. In addition to all the mandatory content, professionals must be educated to know how to educate; above all, they must know how to be a traffic educator and not just act in traffic, attending strategic locations and seeking to improve the Driver/Traffic Officer relationship. It's up to the driver to observe, learn and apply in their daily practice everything that the traffic officer has taught them.

CHAPTER 8

REFERENCES

ANDRADE, L. M. et al. Motorcycle accidents: characteristics of victims and accidents in a hospital in Fortaleza - CE, Brazil. **Rev. Rene.** Fortaleza, v. 10, n. 4, p. 52-59, oct .2009. Available at : <http://www.revistarene.ufc.br/vol10n4_html_site/a06v10n4.htm>. Accessed on 21 November 2010.

ANJOS, K. C. et al. Patient victim of traffic violence: analysis of socioeconomic profile, characteristics of the accident and intervention of the Social Service in the emergency department. **Acta ortop. bras.,** São Paulo, v. 15, n. 5,2007. Assessed at: <http://www.scielo.br/scielo.php?script=sci_arttext&pid =S1413-785220070005 00006&lng=en&nrm=iso>. Accessed on: 16 May 2012.

BACCHIERI, G.; BARROS, A. J. D. Traffic accidents in Brazil from 1998 to 2010: many changes and few effects. **Rev. Saúde Pública,** São Paulo, v. 45, n. 5, Oct. 2011. Available from <http://www.scielo.br/scielo.php?script=sci_ arttext&pid=S0034-89102011000500017&lng=en&nrm=iso>. Accessed on: 18 May 2012.

BENINCASA, M.; REZENDE, M. M. Perception of risk and protective factors for traffic accidents among adolescents. **Boi. Psicol.,** São Paulo, v. 56, n. 125, dec. 2006. Available at: <http://pepsic.bvsalud.org/ scielo.php?script=sci_arttext&pid=S0006-59432006000200008&lng=pt&nrm= iso>. Accessed on: 17 May 2012.

CAIXETA, C.R. et al. Morbidity due to transport accidents among young people in Goiânia, Goiás. **Ciênc. Saúde Coletiva,** v.15, n. 4, p. 2075-2084, 2010. Available at: <http://www.scielo.br/scielo.php?script=sci_arttext&pid =S1413- 81232010000400021 &lng=en&nrm=iso>. Accessed on: 02 February 2011.

CAMPOS, A. K.; BARRAL E. M. R.; DOS SANTOS S. J.; ZUMIOTTI, V. A. **Patient victim of traffic violence: analysis of the socioeconomic profile, characteristics of the accident and social service intervention in the emergency room.**

Acta Ortopédica Brasileira, p. 15262-266, 2007. Available at: <http://redalyc.

uaemex.mx/redalyc/src/inicio/ArtPdfRed.jsp?iCve=65715506>. Accessed on: 17 Apr. 2012.

DEBIEUX, P. et al. Locomotor system injuries in motorbike accidents. **Acta ortop. bras.**, São Paulo, v. 18, n. 6, 2010. Available at: <http://www.scielo.br/scielo.php?script=sci_arttext&pid=S1413-785220100 0060 0010&lng=en&nrm=iso>. Accessed on: 18 May 2012.

NATIONAL TRANSIT DEPARTMENT: Vehicle fleet grows 119 per cent in ten years in Brazil, according to Denatran-2010. Available at: <http://www.denatran.gov.br/ultimas/frota.htm>. Accessed on: 27 Jan. 2011.

DE SOUSA, E. C.; ALMEIDA, J. R. S. **Álcool e adolescentes: fatores** de **risco e consequências dessa relação, Limoeiro do Norte, CE, Brasil** 1,set.2008 [online]. Available at: http://www.webartigos.com/articles/9037/1/Alcool-E- Adolescentes-Fatores-De-Riscois-E-ConsequenciasDessaRelação/pagina1.html. Accessed on: 25 August 2010.

CEARÁ TRAFFIC DEPARTMENT. **Safe transit.** Fortaleza-CE, 2010. Available at: <http://www.detran.ce.gov.br/consultas /files/brochures/DIRECAO_DEFENSIVA.pdf>. Accessed on: 02 Feb. 2010.

PARANÁ TRAFFIC DEPARTMENT. **Community and Traffic. Educating for Traffic.** TECNODATA; Curitiba-PR, 2006. 20ª Ed. 30p. Available at: <http://www.educacaotransito.pr.gov.br/arquivos/File/ Comunidade/Educar%20para%20o%20Transito.pdf >. Accessed on: 14 July 2010.

BRAZILIAN INSTITUTE OF GEOGRAPHY AND STATISTICS - 2010. Available at: <http://www.ibge.gov.br/cidadesat/topwindow.htm71>. Accessed: 13 Feb. 2012.

IMPERATORI, G.; LOPES, M. J. M. Intervention strategies in morbidity due to external causes: how do community health agents work? **Saude Soc.**, v.18, n.1, p. 83-94, 2009. Available at: <http://www.scielo.br/scielo.php?script=sci_arttext&pid=S0104-1290200900010 0009&lng=en&nrm=iso>. Accessed on: 11 December 2010.

JOMAR, R. T. et al. Traffic health education for adolescent high school students. **Esc. Anna Nery,** v.15, n.1, p. 186-189, 2011. Available at: <http://www.scielo.br/scielo.php? script=sci_arttext &pid=S1414- 81452011000100026&lng=en&nrm=iso>. Accessed on: 16 March 2011.

DIÁRIO DO NORDESTE. Traffic accidents on the increase. Fortaleza-CE. Available at: <http://diariodonordeste.globo.com/materia.asp?codigo=744161>. Accessed on: 01 December 2010.

LABIAK, V. B. et al. Exposure factors, experience in traffic and previous involvement in traffic accidents among university students of health courses, Ponta Grossa, PR, Brazil. **Saúde Soc.**, v. 17, n. 1, p. 33-43, 2008. Available at: <http://www.scielo.br/scielo. php?script=sci_arttext&pid=S0104-12902008000100004&lng=en&nrm=iso>. Accessed on: 08 July 2010.

LIBERATTI, C. L. B. et al. Use of helmets by victims of motorbike accidents in Londrina, southern Brazil. **Rev. Panam. Salud Publica,** Washington, v. 13, n. 1, jan. 2003. Available at: <http://www.scielosp.org/ scielo.php?script=sci_arttext&pid=S1020-49892003000100005&lng=en&nrm=is o>. Accessed on: 03 Jun. 2012.

MOURA, E. C. et al. Driving motorised vehicles after abusive consumption of alcoholic beverages, Brazil, 2006 to 2009. **Rev. Saúde Pública,** v. 43, n. 5, p. 891- 894, 2009. Available at: <http://www.scielo.br /scielo.php?script=sci_ arttext&pid=S0034-89102009000500021 &lng=en&nrm=iso>. Accessed on: 20 December 2010.

OLIVEIRA, Z. C.; MOTA, E. L. A.; COSTA, M. C. N. Evolution of traffic accidents in a large urban centre, 1991-2000. **Cad. Saúde Pública,** v. 24, n. 2, p. 364-372, 2008. Available at: <http://www.scielo.br/scielo.php? script=sci_arttext&pid=S0102-311X2008000200015&lng=en&nrm=iso>. Accessed on: 15 October 2010.

OLIVEIRA, N. L. B.; SOUSA, R. M. C. Return to productive activity of motorcyclists victims of traffic accidents. **Acta Paul. Enferm.** v. 19, n. 3, pp.

284-289, 2006. Available at: <http://www.scielo.br /scielo.php?script=sci_ arttext&pid=S0103-21002006000300005&lng=en&nrm= iso> . Accessed on: 03 January 2011.

WORLD HEALTH ORGANISATION. World report on the prevention of road traffic injuries. 2008. Available at: < http//www.who.org>. Accessed on: 01 October 2010.

PARREIRA, J. G. et al . Comparative analysis of injuries found in motorcyclists involved in road traffic accidents and victims of other blunt trauma mechanisms. **Rev. Assoe. Med. Bras.,** São Paulo, v. 58, n. 1, Feb. 2012. Available at: <http://www.scielo.br/scielo.php? script=sci_arttext&pid=S0104-42302012000100018&lng=en&nrm=iso>. Accessed on: 02 June 2012.

PAULINO, L. C. Trânsito no Brasil: desafios à efetivação do direito de ir e vir e permanecer vivo. Fortaleza-CE, **Ed Imprece,** vol. 1, P. 66-67, 2010.

PINTO, A. O.; WITT, R. R. Severity of injuries and characteristics of motorcyclists treated at an emergency hospital. **Revista gaúcha de enfermagem.** Porto Alegre, V. 29, n. 3, p. 408-414, jun. 2008. Available at: <http://hdl.handle.net/10183/23604>. Accessed on: 13 May 2012.

SANTOS, A. M. R. et al. Profile of motorbike accident trauma victims treated at a public emergency service. **Cad. Saúde Pública,** v. 24, n. 8, p. 1927-1938, 2008. Available at: <http://www.scielo.br/ scielo.php?script= sci_arttext&pid=S0102311X2008000800021 &lng=en&nrm=is. Accessed on: 07 January 2011.

SILVA, P. H. N. V. et al . Spatial study of motorbike accident mortality in Pernambuco. **Rev. Saúde Pública,** São Paulo, v. 45, n. 2, Apr. 2011. Available at: <http://www.scielosp.org/scielo.php7script =sci_ arttext&pid=S0034-89102011000200020&lng=en&nrm=Iso>. Accessed on: 16 Mar. 2011.

SOUZA, M. F. M. et al. Descriptive and trend analysis of land transport accidents for social policies in Brazil. **Epidemiol. Serv. Saúde,**

Brasília, v. 16, n. 1, p. 33-44, mar. 2007. Available at: <http://portal.saude.gov.br/portal/arquivos/pdf/revista_vol16_n1_corrigido.pdf>. Accessed on: 07 January 2011.

VERONESE, A. M.; OLIVEIRA, D. L. L. C. The risks of traffic accidents from the perspective of moto-boys: subsidies for health promotion. **Cad. Saúde Pública,** Rio de Janeiro, v. 22, n. 12, dec. 2006. Available at: <http://www.scielo.br/scielo.php?script=sci_arttext&pid=S0102-311X20060012 00021 &lng=en&nrm=iso>. Accessed on: 17 May 2012.

VIEIRA, L. J. E. S. et al. Health team reports on educational practices for traffic victims during hospitalisation/rehabilitation in an emergency hospital. **Saude Soc.** vol. 19, n. 1, p. 213-223, 2010. Available at: <http://www.scielo.br/scielo.php?script=sci_arttext&pid= S0104129020100 00100018&lng=en&nrm=iso>. Accessed on: 21 Dec. 2010.

VIEIRA, R. C. A. et al. Epidemiological survey of motorbike accidents treated at a Trauma Reference Centre in Sergipe. **Rev. esc. enferm. USP,** São Paulo, v. 45, n. 6, Dec. 2011. Available at: <http://www.scielo.br/scielo.php?script=sci_arttext&pid=S00806 23420110 00600012&lng=en&nrm=iso>. Accessed on: 18 May 2012.

CHAPTER 9

APPENDICES

APPENDIX A - Socio-demographic questionnaire. **N°**

SOCIO-DEMOGRAPHIC QUESTIONNAIRE
1 - Age: □ From 10 to 13 years old □ From 14 to 17 years old □ 18 to 20 years old
2 - Sex: □ Male □ Female
3 - Schooling: □ Illiterate □ Primary school □ High school □ Higher education
4 - Marital status: □ Single □ Married □ Divorced □ Others
5 - Family income: □ Less than the minimum wage □ Between one and three minimum wages □ Between four and six minimum wages □ More than six minimum wages
6 - Type of transport used for work □ Cars □ Motorbike □ Others □ Not applicable

APPENDIX B - Questionnaire on motorbike use. N°

1 SKILLS (category A)
1.0 - Are you qualified? □ Yes □ No
1.1 - Are you qualified? □ Yes □ No
1.2 - How long have you been qualified? □ From 01 to 12 months □ From 13 to 24 months □ From 25 to 35 months
2 TRAFFIC SAFETY
2.0 - Are you fully familiar with the Brazilian Traffic Code? □ Yes □ No
2.1 - Helmet present when approached? □ Yes □ No
2.2 - Wearing a helmet: □ Always □ Sometimes □ Never
2.3 - Do you practise defensive driving? □ Yes □ No
2.4 - Do you have your motorbike serviced at the indicated times? □ Yes □ No
2.5 - Have you ever been involved in an accident? □ Yes □ No
3 USE OF ALCOHOL
3.0 - Are you fully familiar with Law 11.705/08 ("Dry Law")?

□ Yes □ No
3.1 - Do you use alcoholic drinks when riding a motorbike? □ Yes □ No
3.2 - Do you or have you ever ridden with an intoxicated motorcyclist? □ Yes □ No
4 TRAFFIC EDUCATION
4.1 - Have you ever taken part in a traffic education programme? □ Yes □ No
4.1.1 - If yes, where did it take place? □ At school □ On the street *(Blitz)* □ Others □ Not applicable
4.2 - Is there traffic education in Quixadá? □ Yes □ No
4.3 - Have you ever been approached by a traffic officer about education? □ Yes □ No

APPENDIX C - Informed Consent Form for adolescents

You are being invited to take part in the research project: MOTORCYCLE ACCIDENTS INVOLVING ADOLESCENTS: An epidemiological study in a rural area of north-eastern Brazil; the main objective of which is to investigate the epidemiology of adolescent motorcyclists in a rural area of north-eastern Brazil between January 2008 and July 2010.

We are asking you to take part in an interview during the DEMUTRAN Blitz in Quixadá to collect information about your behaviour in traffic. I guarantee that the research will not cause any form of harm, damage or inconvenience to those who take part. All the information obtained in this study will be kept confidential and your identity will not be revealed. It is worth emphasising that your participation is voluntary and you can stop taking part at any time, without any harm or damage to the subject. I undertake to use the data collected only for research purposes and the results may be published in scientific articles and specialised magazines or scientific meetings and congresses, as well as being presented to DEMUTRAN in Quixadá for assistance in other *Blitz*, always safeguarding your identity.

All participants will be able to receive any clarifications about the research and, once again, will be free not to take part when they do not feel it is convenient.

You can contact me by phone: (88)96100848 or e-mail: fhvbqxda@vahoo.com.br.

The UECE Ethics Committee is available for clarification by telephone: (85)31019890 - Address Av. Parajana, 1700 - Campos do Itaperi - Fortaleza- Ceará.

This form is drawn up in two copies, one for the research subject and the other for the researcher's files.

I, having been informed about the research, agree to take part in it.

Quixadá, 2012

________________________ ________________________

Participant's signature Signature of researcher

APPENDIX D - Informed Consent Form for those responsible for the adolescents

You, the legal guardian of the minor named below, are being informed of the adolescent's participation in the research: MOTORCYCLE ACCIDENTS INVOLVING ADOLESCENTS: An epidemiological study in a rural area in the north-east of Brazil; the main objective of which is to investigate the epidemiology of adolescent motorcyclists in a rural area in the north-east of Brazil between January 2008 and July 2010.

We ask for your cooperation in this research by authorising the teenager to be interviewed during the DEMUTRAN *Blitz* in Quixadá to collect data on their behaviour in traffic. I guarantee that the research will not cause any form of harm, damage or inconvenience to those who take part. All the information obtained in this study will be kept confidential and your identity will not be revealed. It is worth emphasising that participation is voluntary and you can stop taking part at any time, without any harm or damage to you. I undertake to use the data collected only for research purposes and the results may be published in scientific articles and specialised magazines or scientific meetings and congresses, as well as being presented to DEMUTRAN in Quixadá for assistance in other *Blitz*, always safeguarding your identity.

All participants will be able to receive any clarifications about the research and, once again, will be free not to take part when they do not feel it is convenient.

You can contact me by phone: (88)96100848 or e-mail: fhvbqxda@yahoo.com.br.

The UECE Ethics Committee is available for clarification by telephone: (85)31019890 - Address Av. Parajana, 1700 - Campos do Itaperi - Fortaleza- Ceará.

This form is drawn up in two copies, one for the research subject and the other for the researcher's files.

I, , legal guardian of the adolescent , Having been informed about the research, authorise the adolescent's participation.

Quixadá, 2012

Signature of the person responsible

Signature of researcher

APPENDIX E - Trustee Agreement

To DEMUTRAN

The institution: Departamento Municipal de Trânsito da Cidade de Quixadá Established: Rua Rodrigues Júnior, 675 - Centro.

I'm carrying out a research project on MOTORCYCLE ACCIDENTS INVOLVING ADOLESCENTS: An epidemiological study in a rural area in the north-east of Brazil. In this study I intend to investigate the epidemiology of adolescent motorcyclists in a rural area of north-eastern Brazil between January 2008 and July 2010.

I hereby request authorisation to collect data from the institution's archives on motorbike accidents in the municipality of Quixadá between January 2008 and July 2010.

I clarify that:

- The information collected in the archives will only be used for the purposes of the research;
- I would also like to make it clear that the information will remain confidential and that the anonymity of the teenagers will be preserved.

If you have any questions, please contact the researcher responsible: Francisco Hilângelo Vieira Barros, resident at Rua João Maria de Freitas 1219. João XXIII neighbourhood, telephone: (88)96100848.

I assume responsibility for the term before the Quixadá Municipal Traffic Department.

Quixadá,_______de________ de 2012.

Signature of legal representative

Signature of researcher

APPENDIX F - Final Product

INFORMATIVO SEMESTRAL DO TRÂNSITO DA CIDADE DE QUIXADÁ/CE.

Tiragem 001

OBJETIVO

Manter a população atualizada quanto aos principais eventos sobre o trânsito municipal de Quixadá.

EVENTOS

Futuros: Haverá evento educativo (BLITZ) as 16:00 hs na rua Rodrigues Júnior próximo a praça José de Barros.
Passados: Blitz na Avenida Plácido Castelo próximo ao HEMOCE às 20:00 hs: Intensificação quanto ao porte da carteira de habilitação, uso do sinto de segurança, dos EPI's (capacete) e ingestão de álcool.

ESTATÍSTICA

2008 a 2009: Foram registrados 1064 acidentes, sendo 968 (91%) envolvendo motocicletas. Os cinco bairros ou distritos mais prevalentes nos acidentes são: Centro, Campo Novo, São João, Campo Velho, Alto São Francisco e Custódio. A faixa etária mais acometida é entre 17 e 24 anos
Últimos seis meses:

TRÂNSITO SEGURO

IMPRUDÊNCIA: Ocorre quando o condutor deixa de respeitar qualquer norma, procedimento ou técnica que lhe ofereça segurança.
NEGLIGÊNCIA: Ocorre quando o condutor age com desleixo, quer com seu carro, quer com seu próprio bem estar.
IMPERÍCIA: Ocorre quando o condutor é imperito na prática da direção e em todos os conceitos e habilidades que ela envolve.

PRIMEIROS SOCORROS

Acidentes de trânsito podem acontecer com todos. Mas poucos sabem como agir na hora que eles acontecem. As primeiras providências tomadas em um acidente são chamadas Primeiros Socorros. Eles são os procedimentos de emergência que devem ser aplicados a uma pessoa em perigo de vida, visando a manter os sinais vitais e evitando o agravamento do seu estado até que receba assistência especializada.

ENTENDENDO A LEGISLAÇÃO

INFRAÇÃO GRAVÍSSIMA: Dirigir sob influência de álcool ou de qualquer outra substância psicoativa que determine dependência (7 pontos + Multa e Penalidade: R$ 957,70 + Suspensão do direito de dirigir por 12 meses + Retenção do veículo até a apresentação de condutor habilitado e recolhimento do documento de habilitação).

INFRAÇÃO GRAVE: Dirigir um carro em mau estado de conservação (5 pontos + Multa e Penalidade: R$ 127,69 + Retenção do veículo para regularização).

INFRAÇÃO MÉDIA: Usar placas diferentes das autorizadas pelo CONTRAN (4 pontos + Multa e Penalidade: R$ 85,13 + Retenção do veículo para regularização e apreensão das placas irregulares).

INFRAÇÃO LEVE: Conduzir o veículo sem os documentos de porte obrigatório (3 pontos + Multa e Penalidade: R$ 53,20 + Retenção do veículo até a apresentação do documento).

DIREÇÃO DEFENSIVA

A Direção Defensiva é o conjunto de técnicas que tem como finalidade capacitar o condutor a dirigir de modo a evitar acidentes ou diminuir as ocorrências, apesar das condições adversas ou da ação incorretados outros condutores ou pedestres.

ELEMENTOS DA DIREÇÃO DEFENSIVA

CONHECIMENTO das leis, dos riscos a que estamos expostos, das condições do caminho, etc.
ATENÇÃO constante, pois a qualquer momento pode acontecer uma situação difícil.
PREVISÃO do desenvolvimento das condições do trânsito, com bastante antecedência, e dos riscos a que estaremos sujeitos.
DECISÃO, que implica no reconhecimento das alternativas e em saber decidir a tempo aquela que mais nos convém.
HABILIDADE, ou seja, a capacidade de manejar os controles do veículo e executar perfeitamente as manobras necessárias.

Produto originado da dissertação de mestrado apresentado ao Curso de Mestrado Profissional em Saúde da Criança e Adolescente, Universidade Estadual do Ceará.
Autor: Francisco Hildângelo Vieira Barros
Orientador: Dr. João Joaquim Freitas do Amaral

Faixa de pedestre
A segurança está na sua mão.

TELEFONES ÚTEIS

DETRAN: 0800 275 6768
POLÍCIA: 190
BOMBEIRO: 193
DEMUTRAN: (88) 3412 2728

CHAPTER 10
ANNEXES

ANNEX A - Instrument used at the Dr Eudásio Barroso Municipal Hospital in Quixadá.

ESTADO DO CEARÁ
PREFEITURA MUNICIPAL DE QUIXADÁ
DEPARTAMENTO MUNICIPAL DE TRÂNSITO DE QUIXADÁ
Rua: Rodrigues Júnior, 675 – Irajá. Fone: (88) 3412-2728.

ESTATÍSTICAS DE ACIDENTES DE TRÂNSITO

NOME: ____________________

SEXO: ( ) MASCULINO ( ) FEMININO

IDADE: ____________ DATA: ___/___/___

TIPOS DE ACIDENTES DE TRÂNSITO

( ) ATROPELAMENTO POR MOTOCICLETA OU MOTONETA.

( ) ATROPELAMENTO POR VEÍCULO AUTOMOTOR.

( ) COLISÃO DE MOTOCICLETA COM MOTOCICLETA

( ) COLISÃO DE MOTOCICLETA COM VEÍCULO AUTOMOTOR

( ) COLISÃO DE MOTOCICLETA COM OBJETO FIXO.

( ) COLISÃO DE VEÍCULO AUTOMOTOR COM OBJETO FIXO

( ) COLISÃO DE MOTOCICLETA COM ANIMAL

( ) COLISÃO DE VEÍCULO AUTOMOTOR COM ANIMAL

( ) QUEDA DE MOTOCICLETA

( ) COLISÃO DE MOTOCICLETA COM BICICLETA

( ) OUTROS/ESPECIFICAR____________________

LOCAL ONDE OCORREU O ACIDENTE DE TRÂNSITO:

____

____

ATENDENTE

ANNEX B - Opinion of the Research Ethics Committee

Plataforma Brasil - Ministério da Saúde

UNIVERSIDADE ESTADUAL DO CEARÁ - UECE

PROJETO DE PESQUISA

Título: Prevalência dos Acidentes de Motocicleta Envolvendo os Adolescentes de Quixadá

Área Temática:

Pesquisador: Francisco Hilângelo Vieira Barros **Versão:** 1

Instituição: Curso de Enfermagem **CAAE:** 05312612.0.0000.5534

PARECER CONSUBSTANCIADO DO CEP

Número do Parecer: 77159

Data da Relatoria: 30/07/2012

Apresentação do Projeto:

Trata-se de um estudo do tipo retrospectivo e transversal envolvendo todos os adolescentes entre 10 e 20 anos de idade que utilizam da motocicleta como meio de transporte. O tamanho da amostra será calculado com base na estimativa da média populacional de acordo com o programa EPI INFO, tendo um erro amostral de 5%, grau de confiança de 95% e percentual mínimo de 7% utilizando os dados da frota total de motocicletas (9073 motocicletas) de Quixadá (DENATRAN, 2010).
O resultado da pesquisa que sera apresentado ao Curso de Mestrado Profissional em Saúde da Criança e Adolescente do Centro de Ciências da Saúde , da Universidade Estadual do Ceara, como requisito parcial para obtenção do Titulo de Mestre.

Objetivo da Pesquisa:

Avaliar o perfil epidemiológico dos acidentes de motocicletas em adolescentes do município de Quixadá-CE.
Determinar o número de acidentes de motocicletas nos períodos de janeiro de 2008 a julho de 2010;
Avaliar o perfil sócio demográfico dos adolescentes que utilizam motocicletas como meio de transporte;
Descrever os principais tipos de transporte envolvidos nos acidentes de motocicletas;Identificar as principais causas dos acidentes de motocicletas em adolescentes; Mensurar as principais localidades do município onde ocorreram os acidentes de motocicletas para guiar futuras Blitz.

Avaliação dos Riscos e Benefícios:

Serão incluídos no estudo transversal, somente adolescentes e que estejam conduzindo a motocicleta no momento da Blitz e não oferece riscos aos participantes.

Comentários e Considerações sobre a Pesquisa:

Os índices negativos de acidentes podem ser reduzidos por uma simples medida de prevenção, entretanto o mesmo se torna impossível quando é observada a ampla dependência dos múltiplos fatores envolvidos como: educação, capacitação, revisão das leis, medidas educativas, etc.
Trata um estudo interessante e que poderá trazer benefícios a população.

Considerações sobre os Termos de apresentação obrigatória:

O processo apresenta Folha de Rosto, devidamente preenchida e assinada; Consta o termo de Consentimento Livre e Esclarecido e o Termo de Fiel Depositário estão adequados. Apresenta orçamento na ordem de R$ 106,76 (cento e seis reais e setenta e seis centavos) que sera custeado pelo pesquisador e cronograma com previsão para início da coleta de dados previsto para agosto de 2012.

Recomendações:

O estudo e' relevante e sugiro aprovação

Conclusões ou Pendências e Lista de Inadequações:

A pesquisa esta em conformidade com a resolução 196/96.

Situação do Parecer:

Aprovado

Necessita Apreciação da CONEP:

Não

FORTALEZA, 19 de Agosto de 2012

Assinado por:
DIANA CÉLIA SOUSA NUNES PINHEIRO

Printed by Books on Demand GmbH, Norderstedt / Germany